D1558621

The Mud Snail Son

Betty Jean Lifton

The Mud Snail Son

illustrated by Fuku Akino

Atheneum 1971 New York

Library of Congress catalog card number 77-134816
Published simultaneously in Canada by McClelland & Stewart, Ltd.
Printed in Japan by Toppan Printing Co., Ltd.
Bound by H. Wolff, New York
First Edition

for Carol—
when I was a tadpole
and you were a snail

Once upon a time there was a poor Japanese farmer and his wife who worked from dawn till dusk in the fields of the rich man of their village. Their life was hard because they had to give this greedy landlord most of their rice crops. Still, the only thing they yearned for was a child.

Sometimes when the couple were weeding the water-filled paddies, they would pray to the god of water.

"Oh, please give us a child. We don't care what it is. Even a frog or a mud snail will do as long as it is our very own."

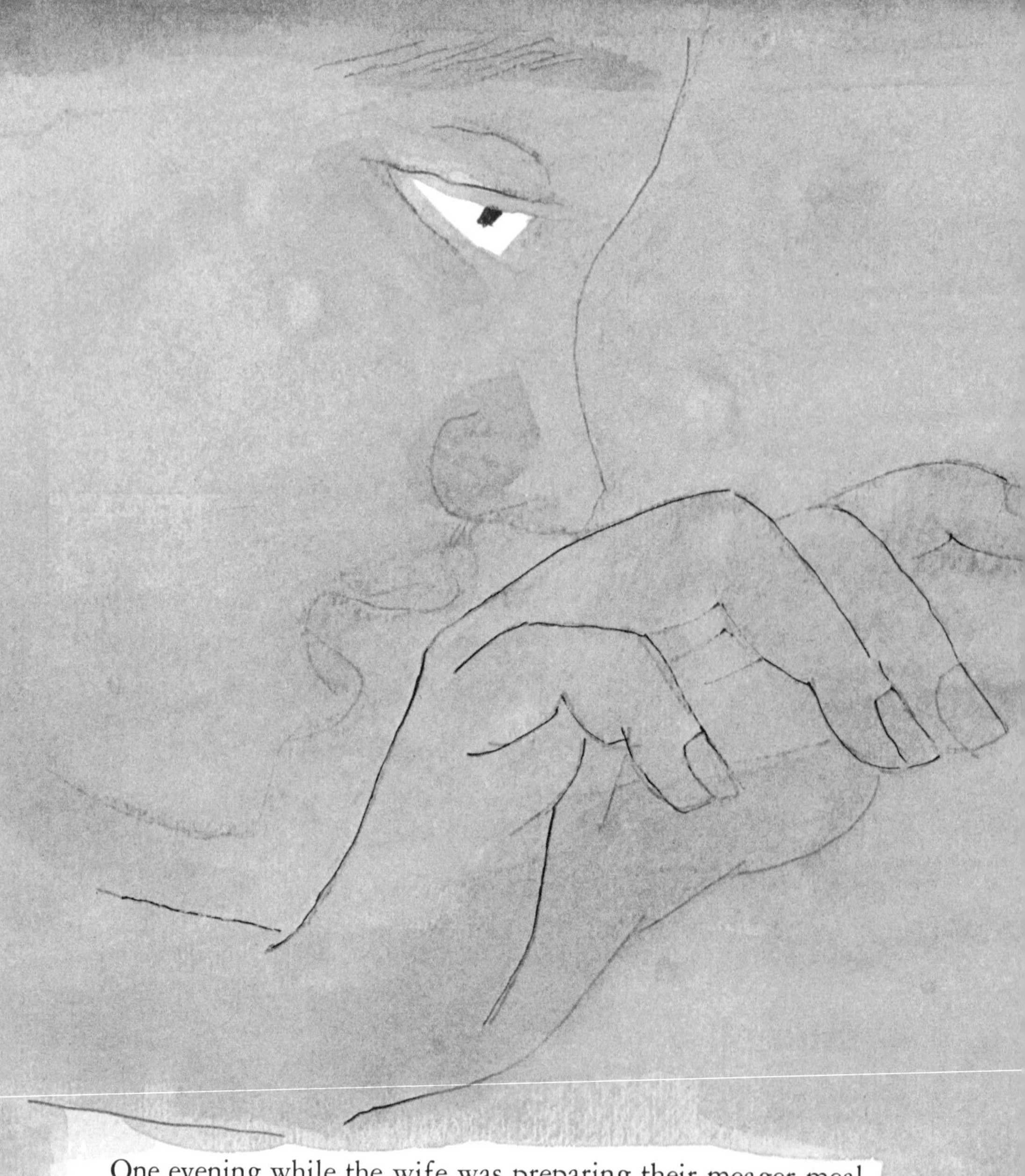

One evening while the wife was preparing their meager meal of rice gruel, she felt very strange. She took to her bed and a few hours later a baby was born—but it was a small mud snail.

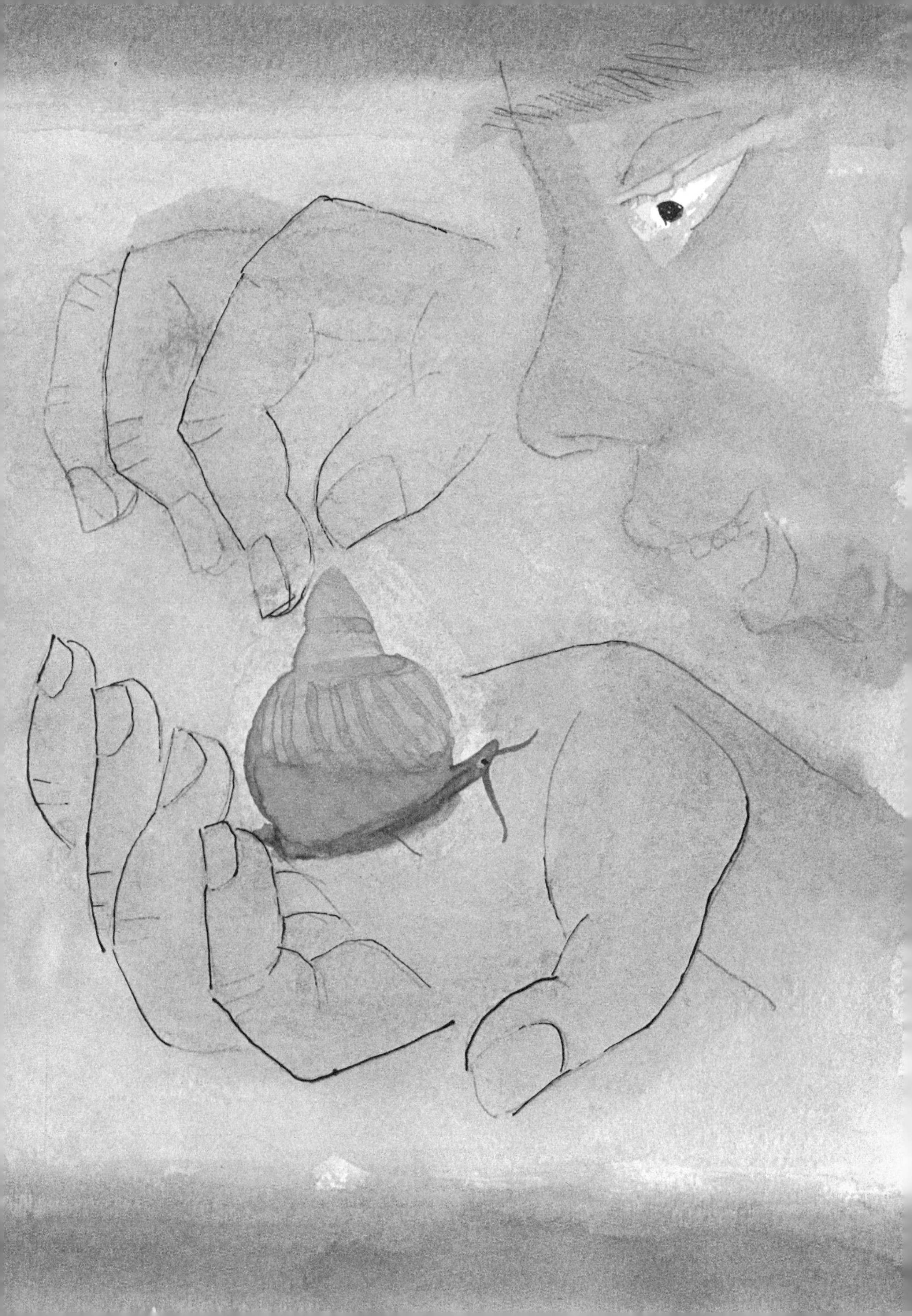

The poor couple were amazed to have a mud snail for a son. But, knowing it was a gift from the water god, they placed it with loving care in a wooden bowl full of water. And they put it high on the narrow shelf of their family shrine so that nothing could get at it. They named it Tany, which is short for "tanishi," meaning mud snail.

Every morning the mother changed her son's water and filled the little dish of rice on his shelf. At night she sang to him, pretending all the while that he was like any other child being tucked into bed. The little mud snail dutifully ate his rice each day, but he did not grow at all. And, though he seemed to listen attentively to his mother's songs, he did not say a word.

Twenty years passed in this way. The farmer and his wife became bent with age. One fall morning after harvest, the old man was so stiff he could hardly load his horse with the bags of rice he had to send to the rich man as land tax.

"I had hoped my son would be able to help me and my wife in our old age," he sighed aloud. "But the god of water sent us only a helpless mud snail for a child. I will have to work like this myself until the end of my days."

And just then a voice said, "Father, I will help you load the rice."

The old man was quite startled and turned around quickly. But he did not see anyone.

"Who is there?" he called nervously.

"It is I, your mud snail son," said the voice from the bowl on the shelf just inside the house. "You and Mother have been very kind to me until now, but it is time for me to take my place in the world. I will deliver the rice to the rich man for you today."

As this was the first time his son had ever spoken to him, the old man was amazed. All he could think to say was, "But how can *you* lead a horse?"

"It will be difficult because I am so small," replied Tany. "Still, if you'll just put me between the rice bags on the horse's back, I will manage."

The farmer thought this was all very peculiar. But then remembering that his son was a gift of the water god, he felt that he should do what he was told. And so calling his wife to come quickly, he picked the mud snail up from the bowl. Together they placed him carefully between the bags of rice on the horse.

"Goodbye, Father! Goodbye, Mother! I am off now," Tany

said in the voice of an ordinary young man. Then calling, "Giddyap!" to the thin horse, he disappeared out the gate.

The old couple were so confused that they hurried to the village shrine to pray for their son's safe return.

"Oh, god of water," they cried, "we have reared the mud snail you gave us as best we know how. Please protect him as he takes our rice to the rich man's house today."

Meanwhile, Tany was guiding the horse perfectly. He even sang a gay packhorseman's song in a deep throaty voice. The horse kept pace with the melody so that the bells around his neck jangled as he pranced along. People on the road and in the fields were astonished to see a riderless horse going by.

"That is the old farmer's lean hag," they told each other. "But who is that singing? Surely it must be a voice without a body."

Finally, Tany arrived at the rich man's house. It was a huge mansion surrounded by rock gardens and high stone walls.

The servants rushed out at the sound of his approach. But they were surprised to see no one on the horse.

"Why should the farmer send his horse here alone?" they were asking each other when a voice called out: "Can someone help me unload the rice?"

The servants turned to the spot where the voice came from, but they saw only a small mud snail sitting between the rice bags.

It said, "Since I am not large enough to do the job myself, would you please take the bales down for me? But first, kindly set me on the veranda of the house so that I do not get crushed."

The servants were so astonished, they ran into the house shouting, "Master! Master! A mud snail has brought your rice."

The rich man, thinking they had gone quite mad, came out

to see for himself. And there indeed was a mud snail, just as they had said. He watched dumbfounded as it directed the servants how to handle the heavy bales, where to place them in the storehouse and how to feed the horse.

The landlord now recalled that long ago the poor farmer and his wife had been given a mud snail by the god of water. But no one had told him it could talk like a human being and

be so clever. Being rather crafty himself, he ordered his wife to prepare a large feast immediately for their unexpected guest.

Tany could hardly be seen as he sat in the place of honor on the low lacquer table. Slowly his bowl of soup disappeared, as if by itself, and then the fish and seaweed and rice. At last his voice boomed out from among the empty dishes, "Everything was delicious. But may I have a cup of water instead of tea?"

"Anything you desire!" cried his host, clapping his hands for water. The rich man could not take his eyes off this miraculous mud snail. Never had he seen one like it in his rice paddies. To have such a creature as this in his family would be like owning a priceless jewel.

While his guest was relaxing, he leaned over the bowl of water and said, "Mud snail, your family and mine have been friendly for many generations. I have two daughters, and I would like to give you one of them for a bride."

Tany was silent for a moment. "Do you really mean it?" he asked finally.

"You will see how much I mean it," said the rich man. And he piled so many presents upon the farmer's thin horse that it could hardly move under the load. Then he sent along two more horses of his own laden with gifts.

The old farmer and his wife had feared the worst when their son had not returned after sunset. They were speechless when he arrived home safely with so many gifts. Still they thought there must be some mistake when he said he was to marry one one of the rich man's daughters.

"Our son has been sent to us by the god of water," they had to keep telling each other. "Perhaps anything is possible."

The next morning the rich man called his two daughters to him and said, "I have promised the mud snail a wife. Which one of you consents to marry him?"

The older daughter was tall and thin, with a voice as sharp as her nose.

"Who would marry a mud snail?" she cried, "I would rather take a worm to be my husband!" And she stalked haughtily from the room.

However, the younger daughter who always tried to help others, said, "Do not worry, my dear father. Since you have promised the mud snail a wife, I will marry him."

This daughter's name was Haru, which means spring. She was gentle as a spring breeze, and her beauty was as fresh and delicate as blossoms that are just unfolding. Her father had hoped to marry her to a handsome prince, but now he could not give up his dream of having the mud snail as a son-in-law.

On the day of the wedding, the rich man gave Haru such a large dowry it was impossible to get it all on seven horses. There were seven trunks full of silk kimonos, and seven trunks of the softest bedding. Altogether, there were so many other treasures that they would not fit in the poor farmer's small house. The landlord had to build him a special storehouse to hold them all.

The rich man invited all his grand friends to the ceremony at his house. They arrived in rustling silks and brocades, each trying to outdo the others in elegance.

But the mud snail invited only one toothless old woman who was a distant relative of his mother. Her kimono, like his parents', was faded and worn. Still everyone agreed it was a very fine, if most unusual affair.

The bride proved to be a joy both to her husband and to the old couple. She waited on Tany with devotion, preparing all his food herself. She even rose before dawn to gather fresh persimmon and taro leaves so that he would have something to nibble on when he awoke. And on the rare days when he wished to go outside, she accompanied him so that birds and animals could not attack him. In truth, Haru only left her husband's side to help his parents in the fields.

Now about half-a-year later, it came time for the special spring festival at the village shrine to give thanks to the water god for the new crops. The old couple left early that day to help with the preparations. Haru cleaned the house, and only when she was satisfied that all the chores were done, did she put on one of her silk kimonos. Tany thought she looked more exquisite than a heavenly nymph.

But just as she was about to go, she suddenly felt sorry to leave her husband alone.

"Why don't you come with me?" she asked.

"I think I will," he said. "The weather is so mild today. And I haven't seen the countryside for a while."

And so the bride put her husband on the knot of the sash at her waist, and they started out together.

Soon they were busy gossiping about the people they saw on the road—about the funny way they walked and talked. Sometimes Haru laughed aloud at the mud snail's sense of humor.

The people passing by and seeing her alone, would shake their heads. "Look, such a beautiful girl laughing and talking to herself," they would whisper. "She's quite mad."

But Haru was having such a good time with Tany, she did not even notice the glances everyone gave her.

There were great crowds at the festival. The mud snail, who was not used to such large gatherings, said, "Why don't you just put me down in a corner of the rice paddy and go into the shrine yourself."

"Are you sure?" Haru asked in surprise.

"It is better that way," he insisted.

Being an obedient wife, she did as she was told and placed him on a muddy ridge in the paddy.

"Be careful not to be found by a crow," she called gaily. "I will just pray to the water god quickly and be right back."

As Haru wandered through the crowd alone, she couldn't help noticing other young women, not nearly so pretty as herself, strolling with their husbands. For a moment she was filled with self pity. When she came to the shrine, she bowed and gave thanks to the water god. But she couldn't help adding with a sigh, "Oh, I do appreciate my wonderful husband. He is gentle and witty and thinks only of my well being. But I am sure he would be even more perfect if he could look like other men."

As soon as she had said this, she was overcome with remorse for even thinking such a thing. She rushed back to the rice paddy to find Tany and take him home.

However, when Haru came to the spot where she had left her husband, he was not there. Thinking he might have started off to meet her, she searched the area thoroughly. But no matter how hard she looked, she could not find him

She went running along the ridge above the paddies peering into the muddy water below. As it was April, there were many mud snails about. She picked them up one by one, turning them over and over in her hands, but none of them spoke to her. None was her husband.

By now she was desperate. She tucked up her silk kimono and waded through the paddies, calling:

"Mud snail, my husband,
Where did you go?
Have you been captured
By some wicked crow?"

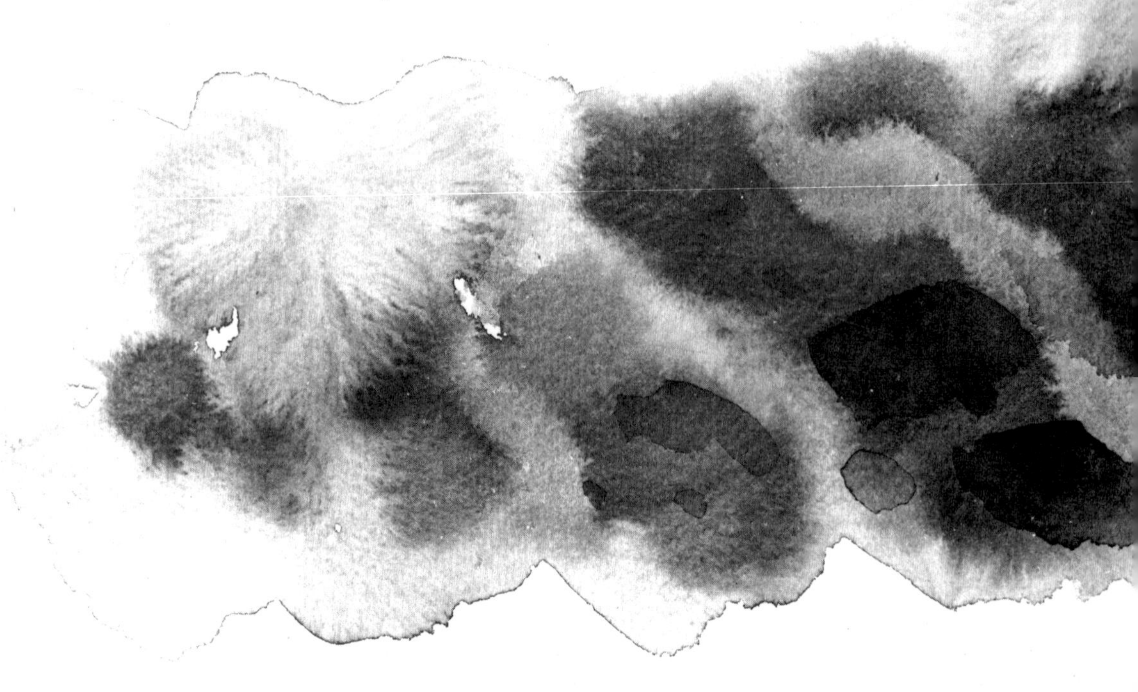

All that day Haru continued picking up every mud snail she saw. Her face and kimono were smeared with mud. And she was so exhausted she kept slipping to the ground. Those who were returning home with their families toward dusk, shook their heads when they saw her.

"What a pity," they told each other. "That lovely girl has obviously lost her mind."

But Haru was not aware of anything except her own misery. "Oh, my dear husband," she was sobbing. "If you'll only come back to me, I will never again wish that you look like anyone else."

As darkness fell over the paddies, Haru decided that she could not return home without Tany. There was just one thing to do—throw herself into the bottomless marsh that lay beyond the shrine. She ran toward it, determined that nothing should stop her. But just as she stood ready to jump from its bank, a voice behind her said gently, "Come now, what are you doing?"

Turning around, she saw a handsome young man smartly dressed with a braided hat on his head, and a bamboo flute in his hand. He looked like a traveller just come to this village.

"Please go away," she wept. "I don't want to go on living without a certain mud snail who was my husband."

"In that case, you have nothing to worry about," he said kindly. "I am your mud snail husband."

At this Haru wept even harder. "I beg you, do not tease me," she pleaded with downcast eyes.

"It is true," said the stranger, leading her away from the marsh. "I am Tany, the child given by the water god to the poor farmer and his wife. Because you loved me, and prayed for me

to become a normal young man, I was able to do so. Please trust me now, and let us go home to tell my parents."

Haru looked up into his eyes. They were so warm and sympathetic, that she knew this must be her husband come back to her in human form. Silently she followed behind him on the road to their house. Before long they were laughing and gossiping together just as they had earlier in the day—as if nothing had happened.

The old couple could hardly believe that this wonderful young man was their son, even after he thanked them for all their care and sang the bedtime songs he remembered hearing as a young mud snail. "Surely the water god works in mysterious ways," they exclaimed.

The rich man was so pleased with his miraculous son-in-law, he built him a grand house and set him up in his own business in the village.

Tany's fame spread rapidly throughout the country. People came from far and wide to see this unusual person for themselves. His business prospered and he became very wealthy.

But Tany never forgot his old mother and father. They lived in great comfort with him and Haru—and they never tired of telling the many visitors to their village about their mud snail son.